Luigi Cornaro

# Sure and Certain Methods of Attaining a Long and Healthy Life

With means of correcting a bad constitution

Luigi Cornaro

**Sure and Certain Methods of Attaining a Long and Healthy Life**
*With means of correcting a bad constitution*

ISBN/EAN: 9783337411374

Printed in Europe, USA, Canada, Australia, Japan

Cover: Foto ©berggeist007 / pixelio.de

More available books at **www.hansebooks.com**

SURE and CERTAIN

# METHODS

OF

ATTAINING

A

## LONG AND HEALTHY LIFE,

WITH MEANS OF CORRECTING A

## BAD CONSTITUTION.

WRITTEN BY

## LEWIS CORNARO,

An Italian Nobleman, when he was near an
HUNDRED Years of Age.

WITH A

*RECOMMENDATORY PREFACE,*

BY THE

HON. JOSEPH ADDISON, ESQ.

𝔗𝔥𝔢 𝔣𝔦𝔯𝔰𝔱 𝔄𝔪𝔢𝔯𝔦𝔠𝔞𝔫 𝔈𝔡𝔦𝔱𝔦𝔬𝔫

———————

PHILADELPHIA,

REPRINTED FOR THE REV. M. L. WEEMS,
BY PARRY HALL, CHESNUT STREET.
M.DCC.XCIII.

# PREFACE.

THE human body is certainly one of the moſt ſtupendous works of Omnipotence. Anatomy diſcovers in it ten thouſand marks of wiſdom and goodneſs, which I have no room to mention here ; nor indeed is it poſſible for any finite intelligence to deſcribe the geometrical accuracy with which the Author of nature has formed every part of the fabric. However, as I ſincerely wiſh that all who are honored with theſe curious and wonderfully wrought bodies, may poſſeſs them in ʰ alth and happineſs ; and, as

long

long experience demonftrates,
that this can not be attained
without TEMPERANCE and EX-
ERCISE, I fhall in this paper
give the reader fuch a view of
the ftructure and mechanifm of
his own frame, as will convince
him of the neceffity and importance of thofe virtues.

I confider the body as a fyf-
tem of tubes and glands, or,
(in a more ruftic phrafe) a large
bundle of pipes and ftrainers;
every part of the body, all the
bowels, mufcles, tendons and
ligaments, are compofed by a
conjunction of countlefs numbers of thefe pipes and ftrainers, that is, of arteries, veins,
nerves and glands. Thefe innumerable veffels, difpofed in
proper order, and filled with
fuita a

suitable fluids or juices, are, by divine appointment, to maintain, while life lasts, a continual action and motion.

The stomach and bowels are continually labouring to digest, that is, to grind and reduce the food into a kind of milk, called chyle; this, carried by millions of little pipes into the blood-veffels, is there, by the unceafing motion of the heart and arteries, converted into blood, and circulated throughout all parts of the body, to repair the conftantly wearing folids, to recruit the ever wafting fluids, and to furnifh a feafonable and friendly fupply to the ceafelefs confumption of nature.

From this fketch of the body and its laws, it plainly appears,

that

that two things are principally essential to good health—Sufficient strength of the veffels— and a free circulation of the fluids.

The veffels muft have ftrength fufficient to convert the food into wholefome blood and urge it on with vigour; and the blood muft have a proper confiftence to yield to the action of the veffels and circulate freely. To preferve the body in this natural and healthy ftate, is an important tafk indeed, and which Infinite Wifdom has configned chiefly to temperance and exercife. The one, allows us to take fuch food only as is wholefome and fufficient to fupply the demands of nature; the other gives fuch firmnefs to the

fibres

fibres as to enable them to digeſt and change the food into
fit nouriſhment, and convey it
to the different parts. And
though the component parts of
our bodies are ſo inconceivably
numerous and complicated;
though they are ſo very minute
and delicate, yet ſo wonderful
is the wiſdom and goodneſs of
God in the diſpoſition of them,
that they would ſeldom or never
be diſordered, were we but duly
temperate and laborious. And
without a proper regard to thoſe
great duties, the moſt ſovereign
medicines in nature will not
have virtue ſufficient to preſerve
us long in health. There is an
anecdote related by ſome of the
oriental writers, which places
the importance of exerciſe in a

proper

proper point of view. A king who had long languished under an ill habit of body, and had taken abundance of medicines to no purpose, was at length cured by the following method. His physician took an hollow ball of wood and filled it with drugs, after which he closed it up so artfully that nothing appeared. He likewise took a mall, and having hollowed it, he inclosed in it several drugs after the same manner as in the ball itself. He then ordered the king to exercise himself every morning with these instruments, till he should get into a moderate perspiration; when as the story goes, the virtue of the medicaments perspiring thro' the wood, had so good an influence

ence on the king's conftitution, that they cured him of an indifpofition which all the compofitions he had taken inwardly had not been able to remove.

This allegory is finely contrived to fhew us how beneficial bodily labour is to health, and that exercife is the beft phyfic. But there is another grand prefervative of health, I mean temperance, which may be practifed by all ranks and conditions, at any feafon, or in any place, without interruption to bufinefs, expence of money, or lofs of time. Thefe two remedies, duly obferved, will fortify the conftitution and render it, in fome fort, invulnerable. Exercife tends to throw off all fuperfluities, and temperance

to

to prevent them; exercife clears
the veffels, temperance never
overftrains them; exercife rolls
on the vital current, temperance
gives nature full play, and allows her to exert herfelf in all
her force and vigour; exercife
diffipates a growing diftemper,
temperance ftarves it.

Phyfic is hardly any thing
elfe but the fubftitute of temperance or exercife. 'Tis indeed
abfolutely neceffary in fome diftempers, but did men but live
in an habitual ufe of thofe two
great inftruments of health,
there would be but little occafion for it. Bliftering, cupping
and bleeding, emetics, diet-
drinks and bitters, are feldom
of ufe but to the intemperate
and idle, who ufe them in order
to

to make their luxury confiftent with health. The apothecary and doctor are perpetually employed in countermining the cook and diftiller. It is faid of Diogenes, that meeting a young man who was going to a feaft, he took him up in the ftreet, ·and carried him home to his friends, as one who was running into imminent danger, had not he prevented him. What would that philofopher have faid, had he been prefent at the gluttony of a modern meal? Would he not have thought the mafter of a family mad, had he feen him devour fowl, fifh and flefh; fwallow oil and vinegar, wines and fpices; throw down fallads of twenty different herbs, fauces of an hundred ingredients,

confections

confections and fruits of numberlefs fweets and flavors ?
What unnatural motions and
counterferments muft fuch a
medley of intemperance produce
in the body ? For my part,
when I behold a fafhionable
table fet out in all its magnificence, I fancy that I fee gouts
and dropfies, fevers and lethar-
gies, with other innumerable
diftempers, lying in ambufcade
among the difhes.

Were I permitted to pre-
fcribe fuch a kind of temperance as would fuit all perfons,
I would copy the following
rules of a very eminent phyfician.

Make your whole repaft out
of one difh. If you indulge
in a fecond, avoid drinking any
thing

thing ſtrong, till you have fi-
niſhed your meal; and abſtain
from all ſauces, or at leaſt ſuch
as are not the moſt plain and
ſimple. A man would ſeldom
be guilty of gluttony if he attended to theſe few and eaſy
rules, prudently contenting himſelf with one good diſh, he would
not be in danger of exceſs, like
thoſe who indulge their craving
appetites on every thing that is
ſet before them. And by ab-
ſtaining from hot ſauces, and
ſtrong drinks, he would never
feel thoſe falſe appetites which
often betray intemperate people
to load their ſtomachs greatly
to their hurt. And ſince it is
to be feared, that the moſt temperate do ſometimes err a little
on the ſide of exceſs, a man

B would

would do well to mifs a meal
now and then ; it would prove
a great relief to nature, help
her to cleanfe and carry off her
crudities, and give her time to
recover the tones and fprings of
her diftended veffels.   Befides,
abftinence well timed, often
kills a ficknefs in the bud, and
deftroys the firft feeds of an indifpofition.   Several eminent
writers of antiquity tell us, that
Socrates lived in Athens all the
time of that dreadful plague
which fwept off fo many thoufands, and yet he never took
the leaft infection ; which thefe
writers unanimoufly afcribe to
his temperate way of living.

And here I cannot but ob-
ferve, that if we compare the
lives of thofe ancient fages who
were

were fo eminent for their temperance, with the lives of any
fet of kings or great men of the
fame number, we fhould think
they were of two different dates.
For the generality of thofe wife
men were nearer an hundred
than fixty years of age at the
time of their deaths. But the
moft remarkable inftance of the
efficacy of temperance towards
procuring long life, is what we
meet with in a little book pub-
lifhed by Lewis Cornaro the Venetian; which I the rather mention becaufe it is of undoubted
credit, as the late Venetian am-
baffador, who was of the fame
family, attefted more than once
in converfation when he refid-
ed in England. Cornaro, the
author of the little treatife I am
menti-

mentioning was of an infirm
conſtitution till about forty,
when by obſtinately perſiſting
in an exact courſe of temperance, he recovered a perfect
ſtate of health, inſomuch that
at fourſcore he publiſhed his
book, which has been tranſlated
into Engliſh, under the title of
" *Sure and certain methods of at-*
" *taining a long and healthy*
" *life.*" He lived to give a third
or fourth edition of it ; and after having paſſed his hundredth
year, died without pain or agony, like one who falls aſleep.
The treatiſe I mention has been
taken notice of by ſeveral eminent authors, and is written
with ſuch a ſpirit of chearful-
neſs, religion, and good ſenſe,
as are the natural concomitants

of

of temperance and ſobriety. The mixture of the old man in it, is rather a recommendation than a diſcredit to it.

# SURE and CERTAIN
# METHODS
## of
## ATTAINING
### A
## LONG AND HEALTHY LIFE.

---

# CHAP. I.

I HAVE obferved that cuf-
tom has lately introduced into
Italy, two very dangerous evils
—FLATTERY and INTEMPE-
RANCE.

The firft of thefe banifhes
from converfation, all frank-
nefs and plain dealing. And a-
gainft the latter I declare open
. war.

war, as being the moſt fatal e-
nemy of our health.

'Tis an unhappineſs into
which the people of this age
are fallen, that variety of diſhes
is become faſhionable and too
generally preferred to frugality.
And yet the one is the offspring
of divine temperance; whilſt
pride and gluttony are the odious parents of the other. Not-
withſtanding the difference of
their origin, yet prodigality is
now a-days tricked up in the
pompous titles of magnificence,
generoſity and grandeur; whilſt
bleſt frugality is too often brand-
ed as the badge of an avaricious
and ſordid ſpirit.

This error has ſo far ſeduced
us, as to prevail on many to renounce a frugal way of living,
though

though taught by nature, from the earliest ages of the world; and has betrayed us into those excesses which serve only to abridge the number of our days. We are grown old before we have been able to taste the pleasures of being young. And the time which ought to be the summer of our lives is often the beginning of their winter. We soon perceive our strength to fail, and weakness to come on, long before we have attained to the perfection of our nature. On the contrary, temperance promotes and preserves to us the full perfection of our natures. Our youth is lasting, and our manhood attended with a vigor that does not begin to decay till after a great many years,

years. This is fo true, that when men were not addicted to intemperance they had more ftrength and vivacity at four-fcore, than we have at forty.

Oh unhappy Italy! doeft thou not fee, that gluttony and ex-cefs rob thee, every year, of more inhabitants than peftilence, war, and famine could have done? Thy true plagues, are thy numerous luxuries and im-moderate feaftings, in which thy deluded citizens indulge themfelves to an excefs unwor-thy of the rational character and utterly ruinous to their health; for how is it poffible to fupport nature under fuch loads of con-trary and unwholefome foods? Put a ftop to this fatal abufe, for God's fake, for there is not,

I am

I am certain of it, a vice more abominable in the eyes of the divine Majefty, nor any more deftructive. How many have I feen cut off, in the flower of their days by this unhappy cuftom of high feeding! How many excellent friends has gluttony deprived me of, who, but for this accurfed vice, might have been an ornament to the world, an honour to their country, and have afforded me as much joy in their lives, as I now feel concern at their lofs!

In order, therefore, to put a ftop to fo great an evil, I have undertaken this little book, and I attempt it the more readily, as many young gentlemen have requefted it of me, moved thereto by feeing their fathers drop

off

off in the flower of their youth,
and me fo found and hearty at
the age of eighty-one. They
begged me to let them know by
what means I attained to fuch
excellent health and fpirits at
my time of life. I could not
but think their curiofity very
laudable, and was willing to
gratify them, and at the fame
time do fome fervice to my
countrymen, by declaring, in
the firft place, what led me to
renounce intemperance and lead
a temperate life ; fecondly, by
fhewing the rules I obferved;
and thirdly, what unfpeakable
fatisfaction and advantage I de-
rived from it ; whence it may
be very clearly feen how eafy a
thing it is for a wife man to e-
fcape all the curfes of intemperance,

perance, and secure to himself the inestimable felicities of vigorous health and chearful age.

The first thing that led me to embrace a temperate life, was, the many and sore evils which I suffered from the contrary course of living; my constitution was, naturally, weakly and delicate, which ought in reason to have made me more regular and prudent, but being, like most young men, too fond of what is usually called good eating and drinking, I gave the rein to my appetites. In a little time I began to feel the ill effects of such intemperance; for I had scarce attained to my thirty-fifth year, before I was attacked with a complication of disorders, such as, head-achs, a

c                                    sick

sick stomach, cholicky uneasi-
nesses, the gout, rheumatic pains,
lingering fevers, and continual
thirst; and though I was then
but in the middle of my days,
my constitution seemed so entirely ruined that I could hardly hope for any other termina-
tion to my sufferings but death.

The best physicians in Italy
employed all their skill in my
behalf, but to no effect. At
last they told me, very candid-
ly, that there was but one thing
that could afford me a single
ray of hope, but one medicine
that could give a radical cure;
viz. the immediate adoption of
a temperate and regular life.
They added moreover, that,
now, I had no time to lose,
that I must immediately, either
chuse

chufe a regimen or death, and
that if I deferred their advice
much longer, it would be too
late for ever to do it. This was
a home thruft. I could not bear
the thoughts of dying fo foon,
and being convinced of their a-
bilities and experience, I thought
the wifeft courfe I could take,
would be to follow their advice,
how difagreeable foever it might
feem.

I then requefted my phyficians to tell me exactly after
what manner I ought to govern
myfelf? To this they replied,
that I fhould always confider
myfelf as an infirm perfon; eat
nothing but what agreed with
me, and that in fmall quantity.
I then immediately entered on
this new courfe of life, and
with

with fo determined a refolution,
that nothing has been fince able
to divert me from it. In a few
days I perceived that this new
way of living agreed very well
with me ; and in lefs than a
twelve month I had the unfpeakable happinefs to find that
all my late alarming fymptoms
were vanifhed, and that I was
perfectly reftored to health.

No fooner had I began to tafte
the fweets of this new refurrec-
tion, but I made many very
pleafing reflections on the great
advantages of temperance, and
thought within myfelf, " if this
" virtue has had fo divine an
" efficacy, as to cure me of fuch
" grievous diforders, furely it
" will help my bad conftituti_
" on and confirm my health."

I there-

I therefore applied myfelf diligently to difcover what kinds of food were propereft for me. I refolved to try whether thofe difhes that pleafed my tafte, were friendly or hurtful to my health, and whether the proverb be true, which fays, *that what delights the palate muft be good for the ftomach.* I found it to be falfe; and that it juft ferves as an excufe to gluttons who are for indulging themfelves in whatever pleafes their appetites.

I therefore took no more notice of the proverb, but made choice of fuch meats and drinks as agreed with my conftitution, and made it an inviolable law with myfelf, *always to rife with an appetite to eat more if I*

*pleafed.*

*pleafed.* In a word, I entirely renounced intemperance, and made a vow to continue the remainder of my life under the fame regimen I had obferved: A happy refolution this! the keeping of which entirely cured me of all my infirmities. I never before lived a year together, without falling once at leaft, into fome violent illnefs: but this never happened to me afterwards; on the contrary, I have always been healthy ever fince I was temperate.

I muft not forget here to mention a circumftance of confiderable confequence. I have been telling of a great, and to me, a moft happy change in my way of living. Now all changes, though from the *worft* to the *beft* habits,

habits, are, at firſt, diſagreeable.
I found it ſo; for having long
accuſtomed myſelf to high feeding, I had contracted ſuch a
fondneſs for it, that though I
was daily deſtroying myſelf,
yet did it, at firſt, coſt me ſome
ſtruggle to relinquiſh it. Nature, long uſed to hearty meals,
expected them, and was quite
diſſatisfied with my moderate
repaſts. To divert my mind from
theſe little diſſatisfactions, I uſed immediately after dinner, to
betake myſelf to ſome innocent
amuſement or uſeful purſuit,
ſuch as, my devotions, my book,
muſic, &c.

But to return.—Beſides the
two foregoing important rules
about eating and drinking, that
is, not to take of any thing, but

**as**

as much as my ſtomach could eaſily digeſt, and to uſe thoſe things only which agreed with me; I have very carefully avoided all *extremes* of *heat* and *cold*, exceſſive fatigue, interruption of my uſual time of reſt, *late hours*, and *too cloſe* and *intenſe thinking*. All theſe are hurtful; but exceſſive fatigue, either of body or mind, is *eminently* ſo. Too cloſe and intenſe thinking ſtrains the nerves, waſtes the ſpirits, brings on a painful head-ach, loſs of appetite, prevents ſleep, ſours the temper, waſtes the fleſh, and, if long continued, effectually deſtroys the beſt conſtitution. Many an excellent conſtitution has been irrecoverably ruined by a few months only of too cloſe hard

hard ftudy ; and the ill effects of this imprudence, are greatly aggravated by that fedentary life, ftooping pofture, and leaning againft tables, which ftudious people are fo often guilty of, and by which they too frequently bring on themfelves pains of the breaft, and incurable confumptions. I am likewife greatly indebted for the excellent health I enjoy, to that calm and temperate ftate in which I have been careful to keep my paffions.

The influence of the paffions on the nerves, and health of our bodies, is fo great, that none can poffibly be ignorant of it. He therefore who ferioufly wifhes to enjoy good health, muft above all things, learn to con-quer

quer his paffions, and keep them in fubjection to reafon. For let a man be ever fo temperate in diet, or regular in exercife, yet ftill fome unhappy paffion, if indulged to excefs, will prevail over all his regularity, and prevent the good effects of his temperance ; no words, therefore, can adequately exprefs the wifdom of guarding againft an influence fo deftructive. Fear, anger, grief, envy, hatred, malice, revenge and defpair, are known by eternal experience, to weaken the nerves, diforder the circulation, impair digeftion, and often to bring on a long train of hyfterical and hypochondriacal diforders ; and extreme fudden fright, has often occafioned immediate death.

On

On the other hand, mode-
rate joy, and all thofe affections of the mind which partake of its nature, as chearfulnefs, contentment, hope, virtuous and mutual love, and courage in doing good, invigorate
the nerves, give a healthy motion to the fluids, promote per-
fpiration, and affift digeftion ;
but violent anger, (which dif-
fers from madnefs only in du-
ration) throws the whole frame
into tempeft and convulfion,
the countenance blackens, the
eyes glare, the mouth foams,
and in place of the moft gentle
and amiable, it makes a man
the moft frightful and terrible
of all animals. The effects of
this dreadful paffion do not ftop
here ; it never fails to create bi-
lious

lious, inflammatory, convulsive, and sometimes apoplectic diforders, and sudden death.

Solomon was thoroughly fenfible of the deftructive tendencies of ungoverned paffions, and has in many places cautioned us againft them. He emphatically ftyles " envy a rot-" tennefs of the bones ;" and fays that, " wrath flayeth the " angry man, and envy killeth " the filly one *;" and " that " the

* The reader will I hope excufe me for relating the following tragical anecdote, to confirm what the benevolent Cornaro has faid on the baneful effects of envy, &c.

In the city of York (England) there died fome time ago, a young lady by the name of D—n. For five years before her death, fhe appeared to be lingering and melancholy. Her flefh withered away, her appetite decayed, her ftrength failed, her feet could no longer fuftain her tottering emaciated body, and her diffolution feemed at hand. One day fhe called her intimate friends to her bed-fide, and as well as fhe could, fpike to the following effect :

" I know

" the wicked ſhall not live out " half their days." For as vi-olent gales of wind will ſoon wreck the ſtrongeſt ſhips, ſo violent paſſions of hatred, an-ger, and ſorrow, will ſoon de-ſtroy the beſt conſtitutions.

D       How-

---

" I know you all pity me, but alas! I am not wor-thy of your pity; for all my miſery is entirely owing to the wickedneſs of my own heart. I have two ſiſters; and I have all my life been unhappy, for no other rea-ſon but becauſe of their proſperity. When we were young, I could neither eat nor ſleep in comfort, if they had either praiſe or pleaſure. As ſoon as they were grown to be women, they married greatly to their ad-vantage and ſatisfaction: this galled me to the heart; and though I had ſeveral good offers, yet thinking them rather unequal to my ſiſters, I refuſed them, and then was inwardly vexed and diſtreſſed, for fear I ſhould get no better. I never wanted for any thing, and might have been very happy, but for this wretched temper. My ſiſters loved me tenderly, for I concealed from them as much as poſſible this odious paſſion, and yet never did any poor wretch lead ſo miſerable a life as I have done, for every bleſſing they enjoyed was a dagger to my heart. 'Tis this Envy, which, preying on my very vitals, has ruined my health, and is now carrying me down to the grave. Pray for me, that God of his infinite mercy may forgive me this horrid ſin; and with my dying breath I conjure you all, to check the firſt riſings of a a paſſion that has proved ſo fatal to me."

However, I muſt 'confeſs to
my ſhame, that I have not been
at all times ſo much of a philo-
ſopher and Chriſtian, as entirely to avoid theſe diſorders; but
I have reaped the benefit of
knowing by my own repeated
experience, that theſe malignant
paſſions have in general a far
leſs pernicious effect on bodies
that are rendered firm and vigorous by temperance, than on
thoſe that are corrupted and weakened by gluttony and exceſs.

That eminent phyſician, Galen, made this obſervation long
before me, and I might produce
ſeveral authorities to ſupport
this opinion, but I will go only
upon my own experience. It
was hard for me to avoid every
extreme of heat and cold, and

to live above all the occasions
of trouble which attend the life
of man ; but yet these things
made no great impreffion on the
ftate of my health, though I
met with many inftances of
perfons who funk under lefs
weight both of body and mind.

'There was in our family a
confiderable law-fuit depend-
ing againft fome perfons, whofe
might overcame our right. One
of my brothers, and fome of my
relations, were fo mortified and
grieved on account of the lofs
of this fuit, that they actually
died of broken hearts. I was
as fenfible as they could be, of
the great injuftice done us, but,
thank GOD, fo far from break-
ing my heart, it fcarcely broke
my repofe. And I afcribe *their*
fufferings

sufferings and *my* safety, to the difference of our living. Intemperance and sloth had so weakened their nerves, and broken their spirits, that they easily sunk under the weight of misfortune. While temperance and active life had so invigorated my constitution, as to make me happily superior to the evils of this momentary life.

At seventy years of age, I had another experiment of the usefulness of my regimen. Some business of consequence calling me into the country, my coach-horses ran away with me; I was overset and dragged a long way before they could stop the horses. They took me out of the coach, with my head batter'd, a leg and an arm out of joint,

joint, and truly in a very lament-
able condition. As soon as they
had brought me home, they
sent for the physicians, who did
not expect I could live three
days : however, they resolved
upon letting me blood, to prevent the fever, which usually
happens in such cases. I was
so confident, that my regular
life had prevented the contract-
ing of any ill humours, that I
opposed their prescription. I
ordered them to dress my head,
to set my leg and arm, to rub
me with some specific oils proper for bruises ; and, without
any other remedies, I was soon
cured, to the great astonishment
of the physicians, and of all
those who knew me.

I beg

I beg leave to relate one more anecdote, as an additional proof what an impenetrable fhield temperance prefents againft the evils of life.

About five years ago, I was over-perfuaded to a thing, which had like to have coft me dear. My relations, whom I love, and who have a real tendernefs for me ; my friends, with whom I was willing to comply in any thing that was reafonable; laft-ly, my phyficians, who were looked upon as the oracles of health, did all agree, that I eat too little; that the nourifhment I took was not fufficient for one of my years ; that I ought not only to fupport nature, but like-wife to increafe the vigor of it, by eating a little more than I did.

did. It was in vain for me to reprefent to them, that nature is content with a little ; that with this little I had enjoyed excellent health fo many years ; that to me the habit of it was become a fecond nature ; and that it was more agreeable to reafon, that as I advanced in years and loft my ftrength, I fhould rather *leſſen* than *increaſe* the quantity of my food, efpeci- ally as the powers of the ftomach muft grow weaker from year to year. To ftrengthen my ar- guments, I urged thofe two na- tural and true proverbs ; one, that he who would eat a great deal muft eat but little ; that is, eating little makes a man live long, and living long he muft eat a great deal. The other

pro-

proverb was, that what we leave, after making a hearty meal, does us more good than what we have eaten. But neither my proverbs nor arguments could silence their affectionate intreaties. Wherefore to please persons who were so dear to me, I consented to increase the quantity of food, but with two ounces only. So that, as before I had always taken but twelve ounces of solid food in the day, I now increased it to fourteen, and as before I drank but fourteen ounces of wine in the day, I now increased it to sixteen. This increase had in eight days time such an effect on me, that from being remarkably chearful and brisk, I began to be peevish and melancholy, and was con-

ftantly

ſtantly ſo ſtrangely diſpoſed, that I neither knew what to ſay to others, nor what to do with myſelf. On the twelfth day I was attacked with a moſt violent pain in my ſide, which held me twenty two hours, and was followed by a violent fever which continued thirty five days, without giving me a moment's reſpite. However, God be praiſed, I recovered, though in my ſeventy eighth year, and in the coldeſt ſeaſon of a very cold winter, and reduced to a mere ſkeleton, and I am poſitive, that, next to God, I am moſt indebted to temperance, for my recovery. O how great is the evil of intemperance, which could, in a few days bring on me ſo ſevere an illneſs, and how glorious

rious are the virtues of temperance, which could thus bear me
up, and fnatch me from the jaws
of death! Order, my friends,
order is every thing; by order,
the arts are more eafily learnt;
by order, armies are rendered
victorious; by order, families,
cities and kingdoms are raifed
to honour and happinefs; and
order is the grand prefervative
of health and long life; nay, I
cannot help faying it is the only
and true medicine. Hence it
is, that when a difinterefted
phyfician vifits a patient, the
firft thing he prefcribes, is to
live regularly. And when he
takes leave of his patient after
recovery, he advifes him, as he
tenders his health, to lead a regular life. And were a patient,
fo

ſo recovered, to live in that manner, he would hardly ever be ſick again. This we may ſay for a certainty, that would all men but live regularly and temperately, there would not be a tenth of that ſickneſs which now makes ſo many melancholy families, nor any occaſion for a tenth part of thoſe nauſeous medicines, which they are now obliged to ſwallow in order to carry off thoſe bad humours with which they have filled their bodies by over eating and drinking. —To ſay the truth, would every one of us but pay a becoming attention to the quantity and quality of what he eats and drinks, and carefully obſerve the effects it has upon him, he would ſoon become his own phyſician, fician, and indeed the very best he could possibly have, for people's constitutions are as different as their faces; and it is impossible, in many very important instances, for the most skilful physicians to tell a man of observation, what would agree with his constitution so well as he knows himself. I am willing to allow that a physician may be sometimes necessary; since there are some disorders against which no human prudence can provide, and which affect us in such a manner as to deprive us of the power of helping ourselves; it is wrong then *wholly* to rely on nature; recourse should be had to some judicious physician, and in cases of danger, the sooner the better. But for the bare purpose pofe of preferving ourfelves in good health, there needs no better phyfic than a temperate and regular life. It is a fpecific and natural medicine, which preferves the man, how tender foever his conftitution be, and prolongs his life to above a hundred years, fpares him the pain of a violent death, fends him quietly out of the world, when the radical moifture is quite fpent, and which, in fhort, has all the properties that are fancied to be in potable gold, which a great many perfons have fought after in vain.

But alas! moft men fuffer themfelves to be feduced by the charms of a voluptuous life. They have not courage enough to deny their appetites; and be-

E                    ing

ing over-perſuaded by their in-
clinations ſo far, as to think
they cannot give up the gratification of them, without abridg-
ing too much of their pleaſures,
they deviſe arguments to perſuade themſelves, that it is more
eligible to live ten years leſs,
than to be upon the reſtraint,
and deprived of whatever may
gratify their appetites. Alas!
they know not the value of ten
years of healthy life, in an age
when a man may enjoy the full
uſe of his reaſon, and turn all
his wiſdom and experience to
his own, and the advantage of
the world. To inſtance only
in the ſciences. 'Tis certain that
ſome of the moſt valuable books
now extant, were written in
thoſe laſt ten years of their authors

thors lives, which some men pretend to undervalue; let fools and villains undervalue life, the world would lose nothing by them, die when they will. But it is a loss indeed, when *wise* and *good* men drop into the grave; ten years of life to men of that character, might prove an inestimable blessing to their families and country. Is such an one a priest only, in a little time he might become a bishop, and by living ten years longer, might render the most important services to the world by his active dissemination of virtue and piety. Is he the aged parent of a family, then though no longer equal to the toils of younger years, yet by his venerable presence and matured counsels,

counfels, he may contribute more
to the harmony and happinefs
of his children, than all their
labours put together.  And fo
with all others,  whether in
church or ftate, army or navy,
who are advanced in years,
though not equal to the active
exercifes of youth, yet in confequence of their fuperior wifdom and experience, their lives
may be of more fervice to their
country, than the lives of thoufands of citizens.  Some, I
know, are fo unreafonable as
to fay that it is impoffible to
lead fuch a regular life.  To
this I anfwer, Galen, that great
phyfician, led fuch a life, and
advifed others to it as the beft-
phyfic. *Plato, Cicero, Ifocrates,*
and a great many famous men
of

of paſt ages embraced it ; and in our time, Pope *Paul Farneze,* Cardinal *Bembo,* and two of our Doges, *Lando* and *Dorato,* have practiſed it, and thereby arrived to an extreme old age. I might inſtance in others of a meaner extract ; but, having followed this rule myſelf, I think I cannot produce a more convincing proof of its being practicable, and that the greateſt trouble to be met with therein, is the firſt reſolving and entering upon ſuch a courſe of life.

You will tell me that *Plato,* as ſober a man as he was, yet affirmed, that it is difficult for a man in public life to live ſo temperately, being often in the ſervice of the ſtate, expoſed to the badneſs of weather, to the

fatigues

fatigues of travelling, and to eat
whatever he can meet with.
This cannot be denied; but then
I maintain, that thefe things
will never haften a man's death,
provided he accuftoms himfelf
to a frugal way of living. There
is no man, in what condition
foever but may keep from over-
eating; and thereby happily
prevent thofe diftempers that are
caufed by excefs. They who
have the charge of public affairs
committed to their truft, are
more obliged to it than any
others: where there is no glory
to be got for their country, they
ought not to facrifice themfelves:
they fhould preferve themfelves
to ferve it; and if they purfue
my method, it is certain they
would ward off the diftempers
which

which heat and cold and fatigues might bring upon them; or should they be disturbed with them it would be but very lightly.

It may likewise be objected, that if one who is well, is dieted like one that is sick, he will be at a loss about the choice of his diet, when any distemper comes upon him. To this I say, that nature, ever attentive to the preservation of her children, teaches us how we ought to govern ourselves in such a case. She begins by depriving us so entirely of our appetites, that we can eat little or nothing. At that time, whether the sick person has been sober or intemperate, no other food ought to be used, but such as is proper

for

for his condition; such as broth, jellies, cordials, barley-water, &c. When his recovery will permit him to use a more solid nourishment, he must take less than he was used to before his sickness; and notwithstanding the eagerness of his appetite, he must take care of his stomach, till he is perfectly cured. Should he do otherwise, he would overburden nature, and infallibly relapse into the danger he had escaped. But notwithstanding this, I dare aver, that he who leads a sober and regular life, will hardly ever be sick; or but seldom, and for a short time. This way of living preserves us from those bad humours which occasion our infirmities, and by conseconfequence heals us of all thofe diftempers which they occafion. I do not pretend to fay that every body muft eat exactly as little as I do, or abftain from fruit, fifh, and other things from which I abftain, becaufe fuch difhes difagree with me. They who are not difordered by fuch difhes, are under no obligation to abftain from them. But they are under the greateft obligations to feed moderately, even on the moft innocent food, fince an overloaded ftomach cannot digeft.

It fignifies nothing to tell me that there are feveral, who, though they live very irregularly, yet enjoy excellent health and fpirits, and to as advanced an an age, as thofe who live ever fo foberly. For this argument is founded on fuch uncertainty and hazard, and occurs fo feldom, as to look more like a miracle than the regular work of nature. And thofe, who, on the credit of their youth and *conftitution*, will pay any regard to fo idle an objection, may depend on it that they are the betrayers and ruiners of their own health.

And I can confidently and truly affirm, that an old man, even of a bad conftitution, who leads a regular and fober life, is furer of a longer one, than a young man of the beft conftitution who lives diforderly. All therefore who have a mind to live long and healthy, and

and die without ficknefs of body or mind, muft immediately
begin to live temperately, for
fuch a regularity keeps the humours of the body mild and
fweet, and fuffers no grofs fiery
vapours to afcend from the ftomach to the head; hence the
brain of him who lives in that
manner, enjoys fuch a conftant
ferenity, that he is always perfectly mafter of himfelf. Happily freed from the tyranny of
bodily appetites and paffions,
he eafily foars above, to the exalted and delightful contemplation of heavenly objects; by
this means his mind becomes
gradually enlightened with divine truth, and expands itfelf
to the glorious enrapturing
view of the Power, Wifdom,

and

and Goodnefs of the Almighty.—He then defcends to nature, and acknowledges her for the fair daughter of GOD, and views her varied charms with fentiments of admiration, joy, and gratitude, becoming the moſt favoured of all fublunary beings.   He then clearly difcerns, and generoufly laments the wretched fate of thofe, who will not give themfelves the trouble to fubdue their paſſions, and thofe three moſt enfnaring luſts, the luſt of the fleſh, the luſt of honours, and the luſt of riches, which all wife and good men have firmly oppbfed and conquered, when they paſſed through this mortal ſtate ; for knowing fuch paſſions to be inconfiſtent with reafon and happinefs,

pinefs, they at once nobly broke through their fnares, and applied themfelves to virtue and good works, and fo, became men of good and fober lives. And when in procefs of time, and after a long feries of years, they fee the period of their days drawing nigh, they are neither grieved nor alarmed. Full of acknowledgments for the favours already received from GOD, they throw themfelves into the arms of his future mercy. They are not afraid of thofe dreadful punifhments, which they deferve who have fhortened their days by guilty intemperance. They die without complaining, fenfible that they did not come into this world to ftay for ever,

but

but are pilgrims and travellers to a far better. Exulting in this faith, and with hopes big with immortality, they go down to the grave in a good old age, enriched with virtues, and laden with honours.

And they have the greater reaſon not to be dejected at the thought of death, as they know it will not be violent, feveriſh or painful. Their end is calm, and they expire, like a lamp when the oil is ſpent, without convulſion or agony, and ſo they paſs gently away, without pain or ſickneſs, from this earthly and corruptible to that celeſtial and eternal life, whoſe happineſs is the reward of the virtuous.

O holy,

O holy, happy, and thrice bleffed temperance ! how worthy art thou of our higheft efteem ! and how infinitely art thou preferable to an irregular and diforderly life ! Nay, would men but confider the effects and confequences of both, they would immediately fee, that there is as wide a difference between them, as there is betwixt light and darknefs, heaven and hell.

Having thus given the reafons, which made me abandon an intemperate, and embrace a fober life, as alfo the method I obferved, and the great bleffings and advantages I reaped from it, I fhall now direct my difcourfe to thofe, who fuppofe it 'to be no benefit to grow old;

becaufe

becaufe they fancy, that when
a man is paft feventy, his life
is nothing but weaknefs, infir-
mity, and mifery.    But I can
affure thefe gentlemen, they
are mightily miftaken; and that
I find myfelf, old as I am,
(which is much beyond what
they fpeak of) to be in the moft
pleafant and delightfome ftage
of life.

To prove that I have reafon
for what I fay, they need only
enquire how I fpend my time,
what are my ufual employ-
ments; and to hear the teftimony of all thofe that know me.
They unanimoufly teftify, that
the life I lead, is not a dead and
languifhing life, but as happy
a one as can be wifhed for in
this world.

They

They will tell you, that I am ftill fo ftrong at fourfcore and three, as to mount a horfe without any help or advantage of fituation; that I can not only go up a fingle flight of ftairs, but climb a hill from bottom to top, a-foot, and with the greateft eafe; that I am always merry, always pleafed, always in humour; maintaining a happy peace in my own mind, the fweetnefs and ferenity whereof appear at all times in my countenance.

Befides, they know that 'tis in my power to pafs away the time very pleafantly; having nothing to hinder me from tafting all the pleafures of an agreeable fociety, with feveral perfons of parts and worth. When

I am

I am willing to be alone, I read good books, and fometimes fall to writing; feeking always an occafion of being ufeful to the public, and doing fervice to private perfons, as far as poffible. I do all this without the leaft trouble; and in fuch times as I fet apart for thefe employments.

I dwell in a houfe, which befides its being fituated in the pleafanteft part of *Padua*, may be looked on as the moft convenient and agreeable manfion in that city. I there make me apartments proper for the winter and fummer, which ferve as a fhelter to defend me from the extreme heat of the one, and the rigid coldnefs of the other. I walk out in my gardens, along my canals and walks; where I
always

always meet with some little
thing or other to do, which, at
the same time, employs and
amuses me.

I spend the months of *April*,
*May*, *September*, and *October*, at
my country-house, which is the
finest situation imaginable: the
air of it is good, the avenues
neat, the gardens magnificent,
the waters clear and plentiful;
and this seat may well pass for
an inchanted palace.

Sometimes I take a walk to
my *Villa*, all whose streets ter-
minate at a large square; in the
midst of which is a pretty neat
church, and large enough for
the bigness of the parish.

Through this *Villa* runs a ri-
vulet; and the country about it
is enriched with fruitful and
well

well cultivated fields; having
at prefent a confiderable number of inhabitants.    This was
not fo formerly: It was a marfhy place, and the air fo unwholefome, that it was more
proper for frogs and toads, than
for men to dwell in.    But on
my draining off the waters, the
air mended, and people refort-
ed to it fo faft, as to render the
place very populous; fo that I
may, with truth, fay that I have
here dedicated to the LORD, a
church, altars, and hearts to
worfhip him; a circumftance
this, which affords me infinite
fatisfaction as often as I reflect
on it.

It is with great fatisfaction
that I fee the end of a work of.
fuch importance to this STATE,
I mean

I mean that of draining and improving fo many large tracts of uncultivated ground, a work which I never expected to have feen compleated, but, thank God, I have lived to fee it, and was even in perfon in thefe marfhy places, along with the commiflaries, for two months together, during the heats of fummer, without ever finding myfelf the worfe for the fatigues I underwent. Of fuch wonderful efficacy is that temperate life which I conftantly obferve.

If in difcourfing on fo important a fubject as this, it be allowable to fpeak of trifles, I might tell you that at the age of fourfcore and three, a temperate life had preferved me in

that

that fprightlinefs of thought,
and gaiety of humour, as to be
able to compofe a very enter-
taining comedy, highly moral
and inftructive, without fhock-
ing or difgufting the audience;
an evil too generally attending
our comedies, and which it is
the duty, and will be the eternal
honour of the magiftracy to dif-
countenance and fupprefs, fince
nothing has a more fatal ten-
dency to corrupt the morals of
youth, than fuch plays as
abound with wanton allufions,
and wicked fneers and fcoffs on
religion and matrimony.

As an addition to my happinefs, I fee myfelf immortalized
as it were, by the great number
of my defcendants. I meet with,
on my return home, not only
twq

two or three, but eleven grand children, all bleſt with high health, ſweet diſpoſitions, bright parts, and of promiſing hopes. I take a delight in playing with the little pratlers; thoſe who are older I often ſet to ſing and play for me on inſtruments of muſick. — Call you this an infirm crazy old age, as they pretend, who ſay, that a man is but half alive after he is ſeventy ? They may believe me if they pleaſe, but really I would not exchange my ſerene chearful old age, with any of thoſe young men, even of the beſt conſtitution, who give the looſe to their appetites; knowing as I do, that they are thereby ſubjecting themſelves every moment to diſeaſe and death.        I re-

I remember all the follies of
which I was guilty in my
younger days, and am perfectly fenfible of the many and
great dangers, they expofed me
to. I know with what vio-
lence young perfons are carried
away by the heat of their blood.
They prefume on their ftrength,
juft as if they had taken a fure
leafe of their lives : and muft
gratify their appetites whatever
it coft them, without confidering that they thereby feed thofe
ill humours, which do moft affuredly haften the approach of
*ficknefs* and *death;* two evils,
which of all others are the moft
unwelcome and terrible to the
wicked. The firft of thefe,
*ficknefs*, is highly unwelcome,
becaufe it effectually ftops their
career

career after this world's busineſs
and pleaſures, which being their
ſole delight and happineſs, muſt
be inexpreſſibly ſad and morti-
fying.　And the impatience and
gloom of ſickneſs is rendered
tenfold more inſupportable to
them, becauſe it finds them utterly deſtitute of thoſe pious affections, which alone can ſoothe
the ſeverity of ſickneſs and charm
the pangs of pain.　They had
never cultivated an acquaintance
with GOD, nor accuſtomed themſelves to look up to him as to a
merciful Father, who ſends af-
fliction to wean us from this
ſcene of vanity.　They had never, by prayers and good works,
endeavoured to ſecure his friend-
ſhip, or cheriſh that love which
would make his diſpenſations

wel-

welcome. So that unbleſt with theſe divine conſolations, the ſeaſon of ſickneſs muſt be dark and melancholy indeed: and beſides all this, their hearts often ſink within them at the proſpect of DEATH, that ghaſtly king of terrors, who comes to cut them off from all their dear delights in this world, and ſend their unwilling ſouls to ſuffer the pu- niſhment which their own guil- ty conſcience tells them is due to their wicked lives.

But from theſe two evils, ſo dreadful to many, bleſſed be GOD, I have but little to fear; for, as for *death*, I have a joyful hope that that change, come when it may, will be glorioufly for the *better*; and beſides, I truſt that HE whoſe divine voice

I have

I have so long obeyed, will gra-
ciously support and comfort his
aged servant in that trying hour.
And as for *sickness*, I feel but
little apprehension on that account, since by my divine medicine TEMPERANCE, I have
removed all the causes of ill-
ness ; so that I am pretty sure I
shall never be sick, except it be
from some intent of *Divine* mercy, and then I hope I shall bear
it without a murmur, and find
it for my good.  Nay, I have
reason to think that my soul has
so agreeable a dwelling in my
body, finding nothing in it but
peace and harmony between my
reason and senses, that she is
very well pleased with her pre-
sent situation ; so that I trust I
have still a great many years to
live

live in health and spirits, and enjoy this beautiful world, which is indeed beautiful to thofe who know how to make it fo, as I have done, and likewife expect (with GOD's affiftance) to be able to do in the next.

Now fince a regular life is fo happy, and its bleffings fo permanent and great, all I have ftill left to do, (fince I cannot accomplifh my wifhes by force) is to befeech every man of found underftanding to embrace, with open arms, this moft valuable treafure of a long and healthy life; a treafure, which, as it far exceeds all the riches of this world, fo it deferves above all things to be diligently fought after, and carefully preferved. This is that divine fobriety, fo

agree-

agreeable to the Deity, the friend
of nature, the daughter of reafon
and the fifter of all the virtues.
From her, as from their proper
root, fpring life, health, chearfulnefs, induftry, learning, and
all thofe employments worthy
of noble and generous minds.
Excefs, intemperance, fuperfluous humours, fevers, pains,
gouts, dropfies, confumptions,
and the dangers of death, va-
nifh, in her prefence, like clouds
before the fun. She is the beft
friend and fafeft guardian of life;
as well of the rich as of the
poor; of the male as of the fe-
male fex; the old as of the young.
She teaches the rich, modefty;
the poor, frugality; men, con-
tinence; women, chaftity; the
old, how to ward off the attacks

of death; and beftows on youth, firmer and fecurer hopes of life. She preferves the fenfes clear, the body light, the underftanding lively, the foul brifk, the memory tenacious, our motions free, and all our faculties in a pleafing and agreeable harmony.

O moft innocent and divine fobriety! the fole refrefhment of nature, the nurfing mother of life, the true phyfic of foul as well as of body. How ought men to praife thee for thy princely gifts, for thy incomparable bleffings! But as no man is able to write a fufficient panegyric on this rare and excellent virtue, I fhall put an end to this difcourfe, left I fhould be charged with excefs in dwelling fo long on fo pleafing a fubject. Yet

as

as numberlefs things may ftill be faid of it, I leave off, with an intention to fet forth the reft of its praifes at a more conveni- ent opportunity.

---

## CHAP II.

### *The method of correcting a bad Conftitution.*

MY treatife on a temperate life has, thank God, be- gun to anfwer my wifhes, in be- ing of fervice to many perfons of weakly conftitutions, who, after every the leaft excefs, found themfelves greatly indifpofed. Thefe gentlemen, on feeing the foregoing treatife, have imme- diately betaken themfelves to a
regular

regular courfe of living, from which, as their numerous letters to me declare, they have experienced the happieft effects. In like manner, I fhould be glad to be of fervice to thofe who are born with good conftitutions, but prefuming too much upon them lead diforderly lives ; whence it comes to pafs, that on attaining the age of fixty or thereabouts, they are attacked with various difeafes; fome with conftant cholicky pains, the tone of the ftomach and bowels being in a manner deftroyed by long continued excefs ; others are tormented with the gout, fome are oppreffed and drowned under dropfical humours, and others worn away to fkeletons by the agonies of the ftone, hectical

tical coughs, and a thousand o-
ther mortal diseases.

I was born with a very chole-
ric, hasty disposition; flew into
a passion for the least trifle, huff-
ed every body about me, and
was so intolerably disagreable,
that many persons of gentle manners absolutely shunned my com-
pany. On discovering how
great an injury I was doing myself, I at once resolved to make
this vile temper give way to reason. I considered that a man
overcome by passion, must at
times, be no better than a madman, and that the only difference between a passionate and a
madman, is, that the one has
lost his reason for ever, and the
other is deprived of it by fits only; but that in one of these, though
never

never fo fhort, he may do fome deed of cruelty or death, that will ruin his character, and deftroy his peace *for ever*. A fober life, by cooling the fever of the blood, contributed much to cure me of this frenzy ; and I am now become fo moderate, and fo much a mafter of my paffion, that no body could perceive that it. was born with me.

A man may likewife, by temperance and exercife, correct a bad conftitution, and, notwithftanding a delicate habit, may live a long time in good health.
. It is true indeed, the moft temperate may fometimes be indifpofed, but then they have the pleafure to think that it is not the effect of their own vices ; that it will be but moderate

derate in its *degree*, and of short continuance.

Many have said to me, " *How can you, when at a table covered with a dozen delicious dishes—how can you possibly content yourself with one dish, and that the plainest too at the table? It must surely be a great mortification to you, to see so many charming things before you, and yet scarcely taste them.*" This question has frequently been put to me, and with an air of surprize. I confess it has often made me unhappy; for it proves that such persons are got to such a pass, as to look on the gratification of their appetites as the highest happiness, not considering that the mind is properly the man, and that it is in the affections

affections of a virtuous and pious mind, a man is to look for
his trueſt and higheſt happineſs.
When I ſit down, with my
eleven grand children, to a table
covered with various dainties,
of which, for the ſake of a light
eaſy ſtomach, I may not, at
times, chuſe to partake, yet
this is no mortification to me;
on the contrary, I often find
myſelf moſt happy at theſe
times. How can it otherwiſe
than give me great delight
when I think of that goodneſs
of GOD, which bleſſes the earth
with ſuch immenſe ſtores of
good things for the uſe of mankind; and which, over and a-
bove all this goodneſs, has put
me into the way of getting ſuch
an abundance of them for my
dear

dear grand children ; and, befides muſt it not make me very
happy to think that I have got-
ten ſuch a maſtery over myſelf
as never to abuſe any of thoſe
good things, but am perfectly
contented with ſuch a portion
of them as keeps me always in
good health. O what a tri-
umph of joy is this to my heart !
What a ſad thing it is that
young people will not take in-
ſtruction, nor get benefit from
thoſe who are older and wiſer
than themſelves ! I may uſe, in
this matter, the words of the
wiſe man, " I have ſeen all
things that are done under the
ſun." I know the pleaſures
of eating, and I know the joys
of a virtuous mind, and can ſay
from long experience, that the

one excelleth the other as far as light excelleth darknefs; the one are the pleafures of a mere animal, the other thofe of an angel.

Some are fo thoughtlefs as to fay, that they had rather be afflicted twice or thrice a year with the gout, the fciatic, and other chronic diftempers, than deny themfelves the pleafure of eating and drinking to the full of fuch things as they like. Such perfons would do well to confider, that by adopting a temperate and active life, they might foon recover fuch vigour of conftitution, as in a great meafure, if not entirely, to throw off thofe painful difeafes, and live in health and chearfulnefs to a fine old age. Whereas by continu-
ing

ing the imprudent practice of
high living*, they keep up the
feverish heat of the blood, relax
their nerves, and so rivet on
themselves those inflammatory
wasting distempers, which will
soon carry them to their graves.

To this some are ready to re-
ply, that for their part they had
rather eat and drink as they like,
though it should shorten their
lives, that is, "give them a short
life and a merry one." It is
really a surprising and *sad* thing,
to see reasonable creatures, so
ready to swallow the most dangerous absurdities. For how,
in the name of common sense,
can the life of a glutton or a sot
be

---

* I would have it carefully remembered, that those
who have been long *afflicted* with the gout, should con-
sult some very experienced Physician, before they make
any great change from high living to abstemiousness.

be a merry one? If men could
eat to excefs, drink to fillinefs,
and ruft in floth, and after all,
fuffer no other harm than the
abridgement of ten or a dozen
years of life, they might have
fome little excufe for calling it
a merry life, though furely it
could appear fo to none but perfons of a fadly vitiated tafte.
But fince high living does fo certainly tend to opprefs and weak-
en the ftomach, filling the whole
habit with fuperfluous and dif-
tempered humours, head-achs,
difordered ftomach, indigeftion,
difturbed fleep, bad dreams, difagreeable tafte in the mouth in
the morning, lofs of appetite,
eructations, fick ftomach, vomitings, diarrhœas, fevers, rheu-
matifms, gouts, confumptions,
apoplex-

apoplexies, &c. &c. I fay, fince
an intemperate life will affuredly fow in our bodies the feeds
of fuch difeafes as will after a
few fhort years of feverifh pleafure, make life a burden to us,
with what face can any reafonable being call this a merry life?

O facred and moft bountiful
Temperance! how greatly am
I indebted to thee for refcuing
me from fuch fatal delufions;
and for bringing me, through
the divine benediction, to the en-
joyment of fo many felicities,
and which, over and above all
thefe favours conferred on thine
old man, haft fo ftrengthened
his ftomach, that he has now a
better relifh for his dry bread
than he had formerly for the
moft exquifite dainties, fo that,

by

by eating little, my ſtomach is often craving after the manna, which I ſometimes feaſt on with ſo much pleaſure, that I ſhould think I treſpaſſed on the duty of temperance, did I not know that one muſt eat, to ſupport life; and that one cannot uſe a plainer or more natural diet.

My ſpirits are not injured by what I eat, they are only reviv-ed and ſupported by it. I can, immediately on riſing from ta-ble, ſet myſelf to write or ſtu-dy, and never find that this ap-plication, though ſo hurtful to hearty feeders, does me any harm; and, beſides, I never find myſelf drowſy after dinner, as a great many do;—the reaſon is, I feed ſo temperately, as ne-ver to load my ſtomach nor op-preſs prefs my nerves, fo that I am always as light, active, and chearful after meals as before.

O what a difference there is between a temperate and an intemperate life! The one beftows health and long life, the other brings on difeafe and untimely death. O thou vile wicked intemperance, my fworn enemy, who art good for nothing but to murder thofe who follow thee; how many of my deareft friends haft thou robbed me of, in confequence of their not believing me! But thou haft not been able to deftroy me according to thy wicked intent and purpofe. I am ftill alive in fpite of thee, and have attained to fuch an age, as to fee around me eleven dear grand children, all of fine under-

understandings, and amiable dif-
pofitions, all given to learning
and virtue ; all beautiful in their
perfons and lovely in their manners, whom, had I not abandoned thee thou infamous fource
of corruption, I fhould never
have had the pleafure to behold.
Nor fhould I enjoy thofe beautiful and convenient apartments
which I have built from the
ground, with fuch highly im-
proved gardens, as required no
fmall time to attain their prefent
fent perfection.   No, thou accurfed hag, thy nature is to im-
poverifh and deftroy thofe who
follow thee.  How many wretched orphans have I feen embrac-
ing dunghills ; how many mife-
rable mothers, with their help-
lefs infants, crying for bread,
                              while

while their deluded fathers, flaves to thy devouring lufts, were wafting their fubftance in rioting and drunkennefs !

But thou art not content with confuming the fubftance, thou wouldeft deftroy the very families of thofe who are fo mad as to obey thee. The temperate poor man who labours hard all day, can boaft a numerous family of rofy cheeked children, while thy pampered flaves, funk in eafe and luxury, often languifh without an heir to their ample fortunes. But fince thou art fo peftilential a vice, as to poifon and deftroy the greateft part of mankind, I am determined to ufe my utmoft endeavours to extirpate thee, at leaft in part. And I promife myfelf, that my dear grand

grand children will declare eternal war againſt thee, and, following my example, will let the world ſee the bleſſedneſs of a temperate life, and ſo expoſe thee, O cruel intemperance! for what thou really art, a moſt wicked, deſperate, and mortal enemy of the children of men.

It is really a very ſurpriſing and ſad thing, to ſee perſons grown to men's eſtate, and of fine wit, yet unable to govern their appetites, but tamely ſubmitting to be dragged by them into ſuch exceſſes of eating and drinking, as not only to ruin the beſt conſtitutions, and ſhorten their lives, but eclipſe the luſtre of the brighteſt parts, and bury themſelves in utter contempt and uſeleſſneſs. O what promiſing hopes

have

have been shipwrecked, what immortal honours have been sacrificed at the shrine of low sensuality ! Happy, thrice happy, those who have early been inured to habits of self-denial, and taught to consider the gratification of their appetites as the unfailing source of diseases and death. Ye generous parents who long to see your children adorned with virtue, and beloved as the benefactors of their kind ; O teach them the unspeakable worth of self-government. Unsupported by this, every advantage of education and opportunity will avail them but little : though the history of ancient worthies, and the recital of their illustrious deeds, may at times kindle up in their bosoms a
flame

flame of glorious emulation, yet alas! this glow of coveted virtue, this flush of promised honor, is transient as a gleam of winter sunshine; soon overspread and obscured by the dark clouds of sensuality.

---

## CHAP. III.

*A Letter from Signior Lewis Cornaro to the Right Reverend Barbaro, Patriarch of Aquileia.*

My Lord,

THE human understanding must certainly possess something divine in its nature. What thanks do we not owe to the divine goodness, for this wonderful faculty of our minds, whereby

by we can, though never fo dif-
tant from them, indulge the pleafure of feeing and converfing
with thofe we love ! How glorious is this invention of writing,
whereby we can eafily commu-
nicate to our abfent friends,
whatever may afford them pleafure or improvement ! By means
of this moft welcome contrivance,
I fhall now endeavour to enter-
tain you with matters of the
greateft moment. It is true indeed, that what I have to tell
you is no news,—but I never told
it you at the age of *ninety one.*
Is it not a charming thing, that I
am able to tell you, that my
health and ftrength are in fo excellent a ftate, that inftead of di-
minifhing with my age, they
feem to increafe as I grow old ?

I  All

All my acquaintance are furpriſ-
ed at it; but I, who know the
cauſe of this ſingular happineſs,
do every where declare it. I endeavour, as much as in me lies,
to convince all mankind, that a
man may enjoy a paradiſe on earth
even after the age of fourſcore.

Now, my Lord, I muſt tell
you, that within theſe few days
paſt, ſeveral learned Doctors of
this Univerſity came to be in-
formed by me, of the method I
take in my diet, having under-
ſtood that I am ſtill healthful and
ſtrong; that I have my ſenſes
perfect; that my memory, my
heart, my judgment, the tone
of my voice, and my teeth, are
all as ſound as in my youth; that
I write ſeven or eight hours a
day with my hand, and ſpend
the

the reſt of the day in walking out a-foot, and in taking all the innocent pleaſures that are allowed to a virtuous man ; even muſic itſelf, in which I bear my part.

Ah, Sir ! how ſweet a voice would you perceive mine to be, were you to hear me, like another *David*, chant forth the praiſes of GOD to the ſound of my Lyre ! You would certainly be ſurprized and charmed with the harmony which I make. Thoſe gentlemen particularly admired, with what eaſineſs I write on ſubjects that require both judgment and ſpirit.

They told me, that I ought not to be looked on as an old man, ſince all my employments were ſuch as were proper for a
youth,

youth, and did by no means re-
femble the works of men advanced in years ; who are capa-
ble of doing nothing after four-
fcore, but loaded with infirmities and diftempers, are perpetually languifhing in pain.

That if there be any of them
lefs infirm, yet their fenfes are
decayed; their fight and hearing
fails them, their legs tremble,
their hands fhake, they can no
longer walk, nor are they capa-
ble of doing any thing : and
fhould there chance to be one free
from thofe difafters, his memory
decreafes, his fpirits fink, and
his heart fails him; he is not half
fo chearful, pleafant and happy
as I am.

Several phyficians were fo good
as to prognofticate to me, ten
years

years ago, that it was impoſſible
for me to hold out three years
longer: however, I ſtill find myſelf leſs weak than ever, and am
ſtronger this year than any that
went before. This ſort of miracle, and the many favours
which I received from GOD, obliged them to tell me, that I
brought along with me at my
birth, an extraordinary and ſpe-
cial gift of nature ; and for the
proof of their opinion they employed all their rhetoric, and
made ſeveral elegant ſpeeches
on that head. It muſt be ac-
knowledged, my Lord, that
eloquence has a charming force
on the mind of man, ſince it
often perſuades him to believe
that which never was, and never could be. I was very much

pleaſed

pleafed to hear them difcourfe;
and could it be helped, fince
they were men of parts who ha-
rangued at that rate ? But that
which delighted me moft, was
to reflect, that age and experience may render a man wifer
than all the colleges in the world
can.    And it was in truth by
their help, that I knew the error of that notion.    To unde-
ceive thofe gentlemen, and at
the fame time fet them right,
I replied, that their way of argu-
ing was not juft : that the favour I received was no fpecial,
but a general and univerfal one:
that there was no man alive,
but what may have received it
as well as myfelf: that I was
but a man as well as others:
that we have all, (befides our
cxiftence,)

exiftence,) judgment and reafon: that we are all born with
the fame faculties of the foul;
becaufe GOD was pleafed that
we fhould all have thofe advantages above the other creatures,
who have nothing in common
with us, but the ufe of their
fenfes : that the Creator has be-
ftowed on us this reafon, and
judgment to preferve our lives :
that man, when young, being
more fubject to fenfe than reafon, is too apt to give himfelf
up to pleafure; and that when
arrived to thirty or forty years
of age, he ought to confider, that,
if he has been fo imprudent as
to lead, till that time, a diforderly life, 'tis now high time
for him to take up and live temperately, for he ought to remember

ber that though he has hitherto
been held up by the vigour of
youth and a good conftitution,
yet he is now at the noon of life,
and muft bethink himfelf of going down towards the grave,
with a heavy weight of years
on his back, of which his frequent pains and infirmities are
certain forerunners ; and that
therefore, if he has not been fo
happy as to do it already, he
ought now, immediately to
change his courfe of life, efpe-
cially with refpect to the quality and quantity of his food,
as 'tis on that the health and
length of our days do fo greatly
depend. For in truth, my Lord,
'tis impoffible for thofe who
will always gratify their appetites, not to ruin their conftitutions;

tions; and that I might not entirely ruin mine, I devoted myself to a fober life. I mult confefs, it was not without great reluctance that I abandoned my luxurious way of living. I began with praying to God, that he would grant me the gift of Temperance, well knowing that he always hears our prayers with delight. Then, confidering, that when a man is about to undertake any thing of importance, he may greatly ftrengthen himfelf in it, by often looking forward to the great pleafures and advantages that he is to derive from it. Juft as the hufband-man takes comfort under his toils, by reflecting on the fweets of abundance; and as the good chriftian gladdens in the fervice

of

of GOD, when he thinks on the
glory of that fervice, and the e-
ternal joys that await him: fo
I, in like manner, by ferioufly
reflecting on the innumerable
pleafures and bleffings of health,
and befeeching GOD to ftrength-
en me in my good refolutions,
immediately entered on a courfe
of temperance and regularity.
And though it was at firft highly difagreeable, yet I can truly
fay, that in a very little time
the difagreeablenefs vanifhed,
and I came to find great delight
in it.

Now on hearing my argu-
ments, they all agreed that I had
faid nothing but what was reafonable ; nay, the youngeft a-
mong them told me that he was
willing to allow that thefe advantages

vantages might be common to all men, but was afraid, they were seldom attained ; and that I muft be fingularly favoured of Heaven to get above the delights of an eafy life, and embrace one quite contrary to it : that he did not look on it to be impoffible, fince my practice convinced him of the contrary, but however, it feemed to him to be very difficult.

I replied, that it was a fhame to relinquifh a good undertaking on account of the difficulties that might attend it, and that the greater the difficulty, the more glory fhould we acquire: that it is the will of the Creator, that every one fhould attain to a long life, becaufe in his old age, he might be freed from
the

the bitter fruits that were pro-
duced by fenfe, and might enjoy the good effects of his reafon; that when he fhakes hands
with his vices, he is no longer
a flave to the devil, and finds
himfelf in a better condition of
providing for the falvation of
his foul : that GOD, whofe
goodnefs is infinite, has ordain-
ed that the man who comes to
the end of his race, fhould end
his life without any diftemper,
and fo pafs, by a fweet and eafy
death, to a life of immortality
and glory, which I expect. I
hope (faid I to him) to die fing-
ing the praifes of my Creator.
The fad reflection, that we muft
one day ceafe to live, is no dif-
turbance to me, though I eafily
perceive that at my age, that
day

day cannot be far off; nor am I afraid of the terrors of hell, becaufe, bleffed be GOD, I have long ago fhaken hands with my fins, and put my truft in the mercy and merits of the blood of *Jefus Chrift.*

To this my young antagonift had nothing to fay, only that he was refolved to lead a fober life, that he might live and die as happily as I hoped to do; and that though hitherto he had wifhed to be young a long time, yet now he defired to be quickly old, that he might enjoy the pleafures of fuch an admirable age.

The defire I had of giving you, my lord, a long entertainment, as being one with whom I could never be weary, has in-

K          clined

clined me to write this long letter to you, and to add one word more before I conclude.

Some senſual perſons give out, that I have troubled myſelf to no purpoſe, in compoſing a treatiſe concerning temperance, and that I have loſt my time in endeavouring to perſuade men to the practice of that which is impoſſible. Now this ſurprizes me the more, as theſe gentlemen muſt ſee that I had led a temperate life many years before I compoſed this treatiſe, and that I never ſhould have put myſelf to the trouble of compoſing it, had not long experience convinced me, that it is a life which any man may eaſily lead, who really wiſhes to be healthy and happy. And, beſides the evidence of my
own

own experience, I have the satisfaction to hear, that numbers on seeing my treatise have embraced such a life, and enjoyed from it the very same blessings which I enjoy. Hence I conclude, that no man of good sense will pay any regard to so frivolous an objection. The truth is, those gentlemen who make this objection, are so unhappily wedded to the poor pleasure of eating and drinking, that they cannot think of moderating it, and as an excuse for themselves, they choose to talk at this extravagant rate. However, I pity these gentlemen with all my heart, though they deserve for their intemperance, to be tormented with a complication of distempers, and to be the victims of their passions a whole eternity.

# CHAP. IV.

*Of the Birth and Death of Man,*

THAT I may not be defici-
ent in that duty of charity,
which all men owe to one another, or lose one moment of that
pleasure which conscious usefulness of life affords; I again
take up my pen. What I am
going to say will be looked on
as impossible, or incredible; but,
at the same time, nothing is
more certain, nor more worthily
to be admired by all posterity.
I am now ninety-five years of
age, and find myself as healthy
and brisk, as if I were but twen-
ty-five.

What ingratitude should I be
guilty of, did I not return thanks

to the divine Goodneſs, for all his mercies conferred upon me ? Moſt of your old men have ſcarce arrived to ſixty, but they find themſelves loaded with in-firmities : they are melancholy, unhealthful ; always full of the frightful apprehenſions of dy-ing : they tremble day and night, for fear of being within one foot of their graves ; and are ſo ſtrongly poſſeſſed with the dread of it, that it is a hard mat-ter to divert them from that dole-ful thought. Bleſſed be God, I am free from their ills and terrors. It is my opinion, that I ought not to abandon myſelf to that vain fear : this I will make appear by the ſequel ; and will alſo evince, how certain I am of living an hundred years.

K 3 But

But that I may obferve a method in the fubject I am treating of, I will begin with man at his birth, and thence accompany him through every ftage of life, to his grave.

I fay then, that fome are born with fo bad a conftitution, that they live but a few days, months or years.

Others are born well fhapèd and healthful, but of a tender make ; and fome of thefe live ten, twenty, thirty, or forty years, without being able to attain to that period which is called old age.

Others there are, who bring along with them a ftrong conftitution into the world, and they indeed live to old age : but it is generally (as already obferved)

an

an old age of ficknefs and for-
row; for which they are to
thank themfelves; becaufe they
moft unreafonably prefume on
the ftrength of their conftitution; and will not on any account, abate of that hearty feeding which they indulged in their
younger days. Juft as if they
were to be as vigorous at four-
fcore as in the flower of their
youth: nay, they go about to
juftify this their imprudence,
pretending that as we lofe our
health and vigour by growing
old, we fhould endeavour to repair the lofs, by increafing the
quantity of our food, fince it is
by fuftenance that man is preferved.

But in this they are danger-
oufly miftaken; for as the natural

ral heat and ftrength of the ftomach leffens as a man grows in years, he fhould diminifh the quantity of his meat and drink, common prudence requiring that a man fhould proportion his diet to his digeftive powers.

This is a certain truth, that fharp four humours on the ftomach, proceed from a flow imperfect digeftion; and that but little good chyle can be made, when the ftomach is filled with frefh food before it has carried off the former meal.—It cannot therefore be too frequently nor too earneftly recommended, that as the natural heat decays by age, a man ought to abate the quantity of what he eats and drinks; nature requiring but very little for the healthy fupport
of

of the life of man, especially
that of an old man. Would
my aged friends but attend to
this single precept which has
been so signally serviceable to
me, they would not be troubled
with one twentieth of those infirmities which now harrass and
make their lives so miserable.
They would be light, active,
and chearful like me, who am
now near my *hundredth year*.
And those of them who were
born with good constitutions,
might live to the age of one
hundred and twenty. Had I
been blest with a robust constitution, I should in all probabi-
lity, attain the same age. But
as I was born with feeble stam-
ina, I shall not perhaps outlive
an hundred. And this moral
certainty

certainty of living to a great age, is to be fure, a moft pleafing and defirable attainment, and it is the prerogative of none but the temperate. For all thofe who (by immoderate eating and drinking) fill their bodies with grofs humours, can have no reafonable affurance of living a fingle day longer: oppreffed with food and fwoln with fu-perfluous humours, they are in continual danger of violent fits of the cholic, deadly ftrokes of the apoplexy, fatal attacks of the cholera morbus, burning fevers, and many fuch acute and violent difeafes, whereby thoufands are carried to their graves, who a few hours before, looked very hale and hearty. And this moral certainty of long life

is

is built on such good grounds, as seldom ever fail. For, generally speaking, Almighty GOD seems to have settled his works on the sure grounds of natural causes, and temperance is (by divine appointment) the natural cause of health and long life. Hence it is next to impossible, that he who leads a strictly temperate life, should breed any sickness or die of an unnatural death, before he attains to the years to which the natural strength of his constitution was to arrive. I know some persons are so weak as to excuse their wicked intemperance, by saying, that " the race is not always to the swift, nor the battle to the strong," and that therefore, let them eat and drink as
they

they pleafe, they fhall not die till their time comes. How fcandaloufly do thefe men mif-underftand Solomon and abufe truth! How would it ftartle us to hear our friends fay, "that let them fleep and play, as they pleafe, they fhall not be beggars till their time comes."

Solomon does indeed fay that "the race is not always to the fwift, nor the battle to the ftrong;" but he muft be no better than a madman, who thence infers, that it is not *generally* fo. For the invariable and eternal experience of mankind demon-ftrates, that ninety nine times in an hundred, the race is to the fwift and the battle to the ftrong, bread to the induftrious, and health to the temperate.

But

But it is a matter of fact, and not to be denied, that, though temperance has the divine efficacy to secure us from violent difeafe and unnatural death, yet it is not to be fuppofed to make a man immortal. It is impoffible but that time, which effaces all things, fhould likewife deftroy that moft curious workmanfhip of GOD, the human body: but it is man's privilege to end his days by a natural death, that is, without pain and agony, as they will fee me, when the heat and ftrength of nature is quite exhaufted. But I promife myfelf, that day is a pretty comfortable diftance off yet, and I fancy I am not miftaken, becaufe I am ftill healthy and brifk, relifh all I eat, fleep

L                          quietly,

quietly, and find no defect in any of my senses. Besides, all the faculties of my mind are in the highest perfection ; my understanding clear and bright as ever ; my judgment sound ; my memory tenacious ; my spirits good ; and my voice, the first thing that fails others, still so strong and sonorous, that every morning and evening, with my dear grand children around me, I can address my prayers and chant the praises of the Almighty. O, how glorious this life of mine is like to be, replete with all the felicities which man can enjoy on this side of the grave ; and exempt from that sensual brutality which age has enabled my better reason to banish, and therewith all its bitter fruits,

fruits, the extravagant paffions and diftrefsful perturbations of mind. Nor yet can the fears of death find room in my mind, as I have no licenfed fins to cherifh fuch gloomy thoughts: neither can the death of relations and friends give me any other grief than that of the firft movement of nature, which cannot be avoided, but is of no long continuance. Still lefs am I liable to be caft down by the lofs of worldly goods. I look on thefe things as the property of heaven; I can thank him for the loan of fo many comforts, and when his wifdom fees fit to withdraw them, I can look on their departure without murmuring.—This is the happinefs of thofe only, who grow old in the

the ways of temperance and virtue ; a happineſs which ſeldom attends the moſt flouriſhing youth who live in vice. Such are all ſubject to a thouſand diſorders, both of body and mind, from which I am entirely free: on the contrary ; I enjoy a thouſand pleaſures, which are as pure as they are calm.

The firſt of theſe is to do ſervice to my country. O! what a glorious amuſement, in which I find infinite delight, in ſhewing my countrymen how to fortify this our dear city of Venice, in ſo excellent a manner, as to make her a famous republic, a rich and matchleſs city. Another amuſement of mine is that of ſhewing this maid and queen of cities, in what manner ſhe

may

may always abound with pro-
viſions, by manuring untilled
lands, draining marſhes, and
laying under water and thereby
fatning fields, which had all a-
long been barren for want of
moiſture. My third amuſe-
ment is in ſhewing my native
city, how, though already
ſtrong, ſhe may be rendered
much ſtronger; and, though
extremely beautiful, may ſtill
increaſe in beauty; though rich,
may acquire more wealth, and
may be made to enjoy better air,
though her air is excellent.
Theſe three amuſements, all
ariſing from the idea of public
utility, I enjoy in the higheſt
degree. Another very great
comfort I enjoy is, that having
been defrauded when young, of

a cona confiderable eftate, I have made ample amends for that lofs, by dint of thought and induftry, and without the leaft wrong done to any perfon, have doubled my income, fo that I am able not only to provide for my dear grand children, but to educate and affift many poor youth to begin the world. And I cannot help faying, I reflect with more pleafure on what I lay out in that way, than in any other.

Another very confiderable addition to my happinefs is, that what I have written from my own experience, in order to recommend *temperance*, has been of great ufe to numbers, who loudly proclaim their obligations to me for that work, feveral

ral of them having fent me word from foreign parts, that, under God, they are indebted to me for their lives. But that which makes me look on myfelf as one of the happieft of men, is, that I enjoy as it were, two forts of lives; the one terreftrial, which I poffefs in fact; the other celeftial, which I poffefs in thought; and this thought is attended with unutterable delight, being founded on fuch glorious objects, which I am morally fure of obtaining, through the infinite goodnefs and mercy of God. Thus I enjoy this terreftrial life, partly through the beneficent influences of temperance and fobriety, virtues fo pleafing to Heaven; and I enjoy, through cordial love of the fame

divine

divine Majesty, the celestial life, by contemplating so often on the happiness thereof, that I can hardly think of any thing else. And I hold, that dying in the manner I expect, is not really death, but a passage of the soul from this earthly life, to a celestial, immortal, and infinitely perfect existence. And I am so far charmed with the glorious elevation to which I think my soul is designed, that I can no longer stoop to those trifles, which, alas! charm and infatuate too great a part of mankind. The prospect of parting with my favourite enjoyments of this life, gives me but little concern; on the contrary, I thank GOD, I often think of it with secret joy, since by that loss

loſs I am to gain a life incom-
parably more happy.

O! who then would be trou-
bled, were he in my place? what good man, but muſt in-
ſtantly throw off his load of worldly ſorrow, and addreſs his grateful homage to the Author of all this happineſs? Howe-
ver, there is not a man on earth, who may not hope for the like happineſs, if he would but live as I do. For indeed I am no angel, but only a man, a ſer-
vant of GOD, to whom a good and temperate life is ſo pleaſing, that even in this world he great-
ly rewards thoſe who practiſe it.

' And whereas many embrace a holy and contemplative life, teaching and preaching the great truths of religion, which is
*highly*

*highly* commendable, the chief employment of fuch being to lead men to the knowledge and worfhip of GOD. O that they would likewife betake themfelves entirely to a regular and temperate life! They would then be confidered as faints indeed upon earth, as thofe primitive chriftians were, who obferved fo conftant a temperance, and lived fo long. By living like them, to the age of one hundred and twenty, they might make fuch a proficiency in holinefs, and become fo dear to GOD, as to do the greateft honour and fervice to the world; and they would befides, enjoy conftant health and fpirits, and be always happy within themfelves; whereas they are now

too

too often infirm and melancholy. If indeed they are melancholy, becaufe they fee GOD,
(after all his goodnefs) fo
ungratefully requited; or becaufe they fee men (notwithftanding their innumerable obligations to love) yet hating and
grieving each other : fuch melancholy is truly amiable and
divine.

But to be melancholy on any
other account, is, to fpeak the
truth, quite unnatural in good
chriftians ; fuch perfons being
the fervants of GOD and heirs
of immortality ; and it is ftill
more unbecoming the minifters
of religion, who ought to confider themfelves, as of all others,
in the moft important, fervice-
able, and delightful employment.                    I know,

I know, many of thefe gentlemen think that GOD does pur-
pofely bring thefe occafions of
melancholy on them, that they
may in this life do penance for
their former fins; but therein,
as I think, they are much miftaken. I cannot conceive how
GOD, who loves mankind, can
be delighted with their fufferings. He defires that mankind
fhould be happy, both in this
world and the next; he tells us fo
in a thoufand places in his word,
and we actually find that there
is not a man on earth, who does
not feel the good Spirit of GOD,
forbidding and condemning
thofe wicked tempers, which
would rob him of that happinefs.
No; it is the devil and fin which
bring all the evils we fuffer, on

our

our heads, and not GOD, who is our Creator and Father, and defires our happinefs : his commands tend to no other purpofe. And temperance would not be a virtue, if the benefit it does us by preferving us from diftempers, were repugnant to the defigns of GOD in our old age.

In fhort, if all religious people were ftrictly temperate and holy, how beautiful, how glorious a fcene fhould we then behold! Such numbers of venerable old men as would create furprife. How many wife and holy teachers to edify the people by their wholefome preachings and good examples! How many finners might receive benefit by their fervent interceffions! How many bleffings might they

M

they shower upon the earth! and not as now, eating and drinking so intemperately, as to inflame the blood and excite worldly passions, pride, ambition, and concupiscence, soiling the purity of their minds, checking their growth in holiness, and in some unguarded moment, betraying them into sins disgraceful to religion, and ruinous to their peace for life.——Would they but feed temperately, and that chiefly on vegetable food, they would as I do, soon find it the most agreeable, (by the cool temperate humors it affords) the best friend to virtuous improvement, begetting gentle manners, mild affections, purity of thought, heavenly mindedness, quick relish of virtue and delight

light in GOD. This was the life led by the holy fathers of the defart, who fubfifted entirely on wild fruits and roots, drinking nothing but pure water, and yet lived to an extreme old age, in good health and fpirits, and always happy within themfelves. And fo may all in our days live, provided they would but mortify the lufts of a corruptible body, and devote themfelves entirely to the exalted fervice of GOD; for this is indeed the privilege of every faithful chriftian as Jefus Chrift left it, when he came down upon earth to fhed his precious blood, in order to deliver us from the tyrannical fervitude of the devil; and all through his immenfe goodnefs.

To

To conclude, fince length of days abounds with fo many bleffings, and I am fo happy as to have arrived at that ftate, I find myfelf bound (in charity) to give teftimony in favour of it, and folemnly affure all mankind, that I really enjoy a great deal more than what I now mention; and that I have no other motive in writing on this fubject, than to engage them to practife, all their lives, thofe excellent virtues of temperance and fobriety, which will bring them, like me, to a happy old age. And therefore I never ceafe to raife my voice, crying out to you, my friends, may your days be many, that you may long ferve GOD, and be fitter for the glory which he prepares for his children !    APPEN-

# APPENDIX.

## GOLDEN RULES

### OF

## HEALTH,

SELECTED FROM HIPPOCRATES, PLU-
TARCH, AND SEVERAL OTHER EMI-
NENT PHYSICIANS AND
PHILOSOPHERS.

OF all the people on the
face of the earth, the A-
mericans are under the greatest
obligations to live temperately.
Formed for commerce, our country abounds with bays, rivers,
and creeks, the exhalations from
which, give the air a dampnefs
unfriendly to the fprings of life.
To counteract this infelicity
of climate, reafon teaches us to
adopt every meafure that may
give tone and vigour to the con-

M 3                    ftitution,

ſtitution. This precaution, at all times neceſſary, is peculiarly ſo in autumn, for then the body is relaxed by the intenſe heat of the dog-days, the air is filled with noxious vapours from putrid vegetables; Nature herſelf wears a ſickly drooping aſpect; the moſt robuſt feel a diſagreeable wearineſs and foreneſs of their fleſh, a heavineſs and ſluggiſhneſs in motion, quick feveriſh fluſhings, and ſudden chills darting along their nerves, (all plain proofs of a ſickly atmoſphere, and tottering health.) Now, if ever we need the aid of all-invigorating temperance, now keep the ſtomach light and vigorous by moderate feeding, the veins well ſtored with healthy blood, and the

the nerves full braced by man-
ly exercife and comely chearfulnefs. Be choice of your diet,
fruit perfectly ripe, vegetables
thoroughly done, and meats of
the eafieft digeftion, with a glafs
or two of generous wine at each
meal, and all taken in fuch prudent moderation, as not to load
but ftrengthen the conftitution.
For at this criticaljuncture, a fingle act of intemperance, which
would fcarcely be felt in the
wholefome frofts of winter, oft-
en turns the fcale againft nature,
and brings on obftinate indigef-
tions, load at ftomach, lofs of appetite, a furred tongue, yellow-
nefs of the eyes, bitter tafte in
the mouth in the morning, bi-
lious vomitings, agues, fevers,
&c. which in fpite of the beft
medicines,

medicines, often wear a man away to a ghoſt. If bleſſed with a good conſtitution, he *may* perhaps crawl on to *winter*, and get braced up again by her friendly froſts; but if old or infirm, it is likely death will overtake him, before he can reach that city of refuge.

" The giddy practice of throwing aſide our winter clothes too early in the ſpring, and that of expoſing our bodies, when overheated to ſudden cold, has deſtroyed more people, than famine, peſtilence and ſword*." *Sydenham.*

Thoſe

* I ſaw (ſays an American officer) thirteen grenadiers lying dead by a ſpring, in conſequence of drinking too freely of the cold water, while dripping with ſweat in a hard day's march, in ſummer. And many a charming girl, worthy of a tender huſband, has ſunk into the icy embraces of death, by ſuddenly expoſing her delicate
frame,

Thofe who, by any accident, have loft a meal, (fuppofe their dinner) ought not to eat a plentiful fupper; for if they do, it will lie heavy on their ftomach, and they will have a more reftlefs night than if they had both dined and fupped heartily. He therefore who has miffed his dinner, and finds himfelf empty and faint, wearied and chilly, fhould make a light fupper of fome fpoon victuals, rather than of any ftrong folid food."*

*Hippocrates.* He

frame, warm from the ball-room, to the cold air. And fince " the univerfal caufe acts not by partial, but by general laws," many a good foul, with more piety than prudence, turning out quite warm from a crowded preaching into the cold air without cloak or furtout, has gone off in a galloping confumption to that happy world, where pain and ficknefs are unknown. What a melancholy thing it is, that people cannot take care of their fouls, without neglecting their bodies, nor feek their falvation without ruining their health!

. * I have often (fays Doctor Mackenzie) experienced the benefit

He who has taken a larger quantity of food than usual, and feels it heavy and troublesome on his stomach, will, if he is a wise man, go out and puke it up immediately\*. *Hippocrates.*

And here I cannot omit mentioning a very ruinous error into which too many are fond of running, I mean, the frequent use of strong vomits and purgatives. A man every now and then

benefit of this advice, when in the hurry of country practice, I chanced to lose my dinner and return home tired; for if I ate a hearty meat supper, I was sure to be sick, but if I supped on a dish of chocolate, or a bowl of gruel and butter, or toast and wine made weak and spiced, I rested perfectly well, and rose next morning fresh and chearful.

\* The wise son of Sirach confirms this precept, and says, Ecclef. xxxi. 21. " If thou hast been forced to eat, arise, go forth and puke, and thou shalt have rest." And most certain it is, (adds an ingenious phyfician) that hundreds and thousands have brought sickness and death on themselves, by their ignorance or neglect of this rule. But at the same time people should carefully avoid a repetition of that excess, which renders such an evacuation neceffary, for frequent vomitings do greatly tend to weaken and destroy the tone of the stomach.

then feeds too freely on some favourite diſh; by ſuch exceſs the ſtomach is weakened, the body filled with ſuperfluous humours, and he preſently finds himſelf much out of ſorts. The only medicine in this caſe, is moderate exerciſe, innocent amuſement, and a little abſtinence, this is nature's own preſcription, as appears by her taking away his appetite. But having long placed his happineſs in eating and drinking, he cannot think of relinquiſhing a gratification ſo dear to him, and ſo ſets himſelf to force an appetite by drams, ſlings, elixir of vitriol, wine and bitters, pickles, ſauces, &c. and on the credit of this artificial appetite, feeds again as if he poſſeſſed the moſt vigorous health.

health. He now finds himfelf *entirely* diforderd, general hea-vinefs and wearinefs of body, flatulent uneafinefs, frequent eructations, lofs of appetite, dif-turbed flumbers, frightful dreams, bitter tafte in the mouth, &c. He now complains of a foul ftomach, or (in his own words) that his ftomach is full of bile; and immediately takes a dofe of tartar emetic or a ftrong purgative, to cleanfe out his fto-mach, and fo prepare for another courfe of high living. Of all the Apollyons or deftroyers of nerves, health and life, this is the greateft; and I have no fort of doubt on my own mind but it has broken down more con-ftitutions, brought on more dif-tempers and fent more people to

an

an early grave, than all the vices of this bedlam world put together. How much wifer would it be in this cafe to follow the advice of the celebrated Boerhaave, i. e. to ufe a little abftinence, take moderate exercife, and thereby help nature to carry off her crudities and recover her fprings. I have been often told by a lady of quality, whofe circumftances obliged her to be a good œconomift, and whofe prudence and temperance preferved her health and fenfes unimpaired to a great age, that fhe had kept herfelf out of the hands of the phyficians many years by this fimple regimen. People in health fhould never force themfelves to eat when they have no appetite; Nature, the beft judge

N

in

in thefe matters, will never fail to let us know the proper time of refrefhment. To act contrary to this rule, will affuredly weaken the powers of digeftion, impair health and fhorten life. *Plutarch.*

"Let us beware of fuch food as tempts us to eat when we are not hungry, and of fuch liquors as entice us to drink when we are not thirfty." *Socrates.*

"He who was active and nimble before meals, if he becomes heavy and dull after meals, has certainly tranfgreffed the laws of temperance ; for the true end of eating and drinking is to refrefh, and not to opprefs the body." *Leffius.*

It is really furprifing (fays Plutarch) what benefit men of

letters

letters would receive from reading aloud every day; we ought
therefore to make that exercife
familiar to us, but it fhould not
be done immediately after dinner, nor fatigue, for that error
has proved hurtful to many. But
though loud reading is a very
healthy exercife, violent voci-
feration is highly dangerous; it
has in thoufands of inftances
burft the tender blood veffels of
the lungs, and brought on incurable confumptions*.

"The

---

* Would to God, all minifters of religion (I mention
*them* becaufe they are generally moft wanting in this great
article of prudence) would but attend to the advice of
this eminent Philofopher. They would, many of them,
live much longer, and confequently ftand a good chance
to be more ufeful men here on earth, and brighter
faints in heaven. What can give greater pain to a man
who has the profperity of religion at heart, than to fee an
*amiable, pious young divine*, (who promifed great fer-
vices to the world) fpitting up his lungs, and dying of a
confumption brought on by preaching ten times louder
than he had need! Since the world began, no man ever
fpoke

" The world has long made a juſt diſtinction betwixt men of learning, and wiſe men. Men of learning are oft-times the weakeſt of men : they read and meditate inceſſantly, without allowing proper relaxation or re-freſhment to the body; and think that a frail machine can bear fatigue as well as an immortal ſpirit. This puts me in mind of what happened to the camel in the fable; which, refuſing tho' often premoniſhed, to eaſe the ox in due time of a part of his load, was forced at laſt to carry not only the ox's whole load, but the ox himſelf alſo, when he died under his burden. Thus it happens

ſpoke with *half* the energy which the intereſts of eternal ſouls deſerve, but there is a wide difference betwixt an *inſtructive, moving, melting eloquence,* and a *loud, unmeaning monotony.*

happens to the mind which has
no compaſſion on the body,
and will not liſten to its complaints, nor give it any reſt, un-
til ſome ſad diſtemper com-
pels the mind to lay ſtudy and
contemplation aſide ; and to lie
down, with the afflicted body,
upon the bed of languiſhing and
pain. Moſt wiſely, therefore,
does Plato admoniſh us to take
the ſame care of our bodies as
of our minds ; that like a well
matched pair of horſes to a cha-
riot, each may draw his equal
ſhare of weight. And when
the *mind* is moſt intent upon virtue and uſefulneſs, the *body*
ſhould then be moſt cheriſhed by
prudence and temperance, that
ſo it may be fully equal to ſuch
arduous and noble purſuits."—
*Plutarch.*      N 3      Nothing

Nothing is more injurious to health than hard ſtudy at night; it is inverting the order of nature, and ruining the conſtitution.

All who are ſo wiſe as to riſe early, and ſpend the day in uſeful induſtry, will, by the time night's ſable curtains are drawn, feel the need of that balmy reſtorative, *ſleep*. Now when nature is already exhauſted, and needs repoſe, to go to hard ſtudy, what is it but to ſtrain the nerves, waſte the ſpirits, bring tireſome watchfulneſs, loſs of appetite and general diſorder? But moſt of all, is it not improper to lie reading in bed by candle light? for it not only partakes of the uſual inconveniences of night ſtudy, ſuch as ſtraining

ſtraining the eyes, weakening the ſight, fatiguing the mind, and wearing away the conſtitution, but is oft-times the cauſe of the ſaddeſt calamities; thouſands of elegant houſes, with all their coſtly furniture, have been reduced to aſhes by this very imprudent practice. I knew an amiable lady, who was not a little tried by this ill habit in one of her acquaintance. He would ſleep all the morning, play the truant all day, and at night nothing would ſerve him, but he muſt read in bed till midnight, with a blazing candle all the time cloſe to the curtains. The thought of this, as *well it might*, would not allow the lady a moment's reſt, nor a ſingle wink to the maid, who, poor thing! was

packed

packed up ftairs every quarter of an hour, to take a peep at the candle. I fuppofe it might take well nigh the *whole* of an *angel's benevolence* to keep up a twelve month intimacy with fo dangerous and troublefome a vifitant. But admitting this habit of night ftudying and reading in bed, were attended with none of thefe alarming inconveniences ; no wife man would indulge in it, for it is evident he would fave no time, gain no pleafure nor advantage from it. For, it is very certain *we muft fleep*, and the paternal hand of GOD draws over us the fhades of night for that purpofe ; and if we don't fleep *then*, we muft do it in the day, and is it not a thoufand times better to fleep in the night, while darknefs darkness veils from our eyes all the beauties of creation, and unwholesome damps make it dangerous to ftir out, than to fnore in bed all the morning, when the cheerful light, the chirping birds, the fragrant air, and gladdening fight of gay-green landfkips, together with the fpiritftirring voice of glorious toil, invite to health, to ufefulnefs, and pleafure?

But how can giddy youth, hurried on by ftrong paffions and appetites, be prevented from running into thofe exceffes, which may cut them off in the prime of their days, or at leaft hoard up difeafes and remorfe for old age? Why, their paffions and appetites muft early be reftrained by proper difcipline and

and example. This important office muſt be done by their parents, whoſe firſt and greateſt care ſhould be " to train up their children in the way they ſhould go, that when they are old they may not depart from it."

" O that parents (ſays the excellent Mr. Locke) would carefully inſtill into their children that great principle of all virtue and worth, viz. nobly to deny themſelves every wrong deſire, and ſteadily follow what reaſon dictates as beſt, though the appetite ſhould lean the other way. We often ſee parents by humouring them when little, corrupt the principles of virtue in their children; and wonder afterwards to taſte the bitter waters of their undutifulneſs or wickedneſs,

wickedneſs, when they themſelves have contributed thereto.

Why ſhould we wonder, that he
who has been accuſtomed to
have his will in every thing,
when he was in coats, ſhould deſire and contend for it when he
is in breeches ? Youth is the
golden ſeaſon to inure the mind
to the practice of virtue, on
which their future health and
reſpectability depend, and without which it will be impoſſible
to deliver their conſtitutions,
unbroken, to manhood and old
age. Vice is utterly inconſiſt-
ent with health, which can never dwell with lewdneſs, luxury, ſloth and violent paſſions.
The life of the epicure and rake,
is not only ſhort, but miſerable.
It would ſhock the modeſt and
compaſſionate,

compaffionate, to hear of thofe
exquifite pains, and dreadful
agonies, which profligate young
perfons fuffer from their debau-
cheries, before they can even
reach the friendly fhelter of an
untimely *grave*. Or if fome
few ftop fhort in their career of
riot, before they have quite deftroyed the fprings of life, yet
thofe fprings are generally rendered as feeble and crazy, by the
liberties which they have already taken, that they only fupport a gloomy, difpirited, dying life, tedious to themfelves,
and troublefome to all around
them; and (which is ftill more
pitiable) often tranfmit their
complaints to an innocent unhappy offspring."

www.ingramcontent.com/pod-product-compliance
Lightning Source LLC
Chambersburg PA
CBHW021812190326
41518CB00007B/557

www.ingramcontent.com/pod-product-compliance
Lightning Source LLC
Chambersburg PA
CBHW021813190326
41518CB00007B/573